I0511887

No Return:

An Essay on the Impending Human Extinction.

After reading this you will no longer be able to claim ignorance. You will no longer be able to feign feeling peace and security for your lives and the lives of your children. You will cease to be able to exist comfortably in a state of indifference in regard to the past, present, and future of the human species. Your natural instinct to not process the information presented because of the implications it has on your life will be forcibly attacked in the following text. Legitimately have an internal dialogue with yourself about whether or not you should proceed.

The evidence is starting to become overwhelming. Your life and the lives of your offspring are clearly and presently threatened. The planet itself is in danger of becoming uninhabitable a lot sooner than previously thought. The eyes water and the spine shivers at the realization of humanities few available options on a course to take. Options that appear to all end in not just collapse, not only destruction, but extinction.

First to actually understand and process what we collectively as a species are about to be dealing with we have to look at the data and information. It is absolutely crucial to analyze the information and educate ourselves. The data we are about to briefly cover has, as of now on October 9th, 2018, been public for only a few days. When we seek answers to collective problems solutions come from one place. It makes no difference if its health issues, environmental, technological, societal, or universal. Science and the scientific method have been paramount to improving lives and advancing societies. Breakthroughs and solutions originate from the scientific method. That is demonstrable and overwhelming. The solution to these problems does not come from Jesus or Allah or Yaweh. The solutions do not come from wealth. It's time to get honest with ourselves and each other. The time for tiptoeing around complex issues and fundamental human experiences needs to end if we are to have any hope of limping into the future intact.

Let's discuss the data. The Earth is heating up, it just is. The following statement comes from NASA and 17 other scientific associations, including the American Meteorological Society, The American Associations for the Advancement of Science. The statement is this, "Observations throughout the world make it clear that climate change is occurring, and rigorous scientific research demonstrates that the greenhouse gases emitted by human activities are the primary driver."(1)

Now I understand the tendency to make large mental leaps to dismiss information that does not fit with your personal understanding of the reality that we exist in so I will keep providing you with scientific information from a litany of sources. The next statement on the existence of climate change comes from The Guardian speaking to the misinformation surrounding the topic of climate change and the scientific consensus about it. "Four years ago, my colleagues and I published a paper finding a 97% consensus in the peer-reviewed literature on human-caused global warming. Since then, it's been the subject of constant myths, misinformation, and denial. In fact, last year we teamed up with the authors of six other consensus papers, showing that with a variety of different approaches, we all found the expert consensus on human-caused global warming is 90–100%.

Most of the critiques of our paper claim the consensus is somehow below 97%. For example, in a recent congressional hearing, Lamar Smith (R-TX) claimed we had gone wrong by only considering "a small sample of a small sample" of climate studies, and when estimated his preferred way, it's less than 1%. But in a paper published last year, James Powell argued that the expert consensus is actually higher – well over 99%."(2)

The scientific community overwhelmingly agrees climate change is happening and humans are accelerating that process. That's not the real concerning info though. The real sobering and presently un-processable realization for us all to collectively grasp is the rate at which the planet is now heating.

The UN Intergovernmental Panel on Climate Change released a report on Monday, October 1st that rings the alarm so loud one can only shelter themselves from it for so long. Eventually we are going to have to come to grips with what this all means. The report comments on the need to progress toward keeping the global temperature rise below 1.5 degrees Celsius this century in order to stave of cataclysmic implications. The report finds that not only are we not on track to meet the goal but instead we are heading in the opposite direction. If the world continues to operate the way it does, we are speeding toward a 3-degree increase. 3 degrees in under a century means a total breakdown of society and life as we know it. The report says that in order to have a chance of changing the course with an ever-decreasing timeframe society will have to change, dramatically. In an article by the BBC about the report author Matt McGrath said the following, "Keeping to the preferred target of 1.5C above pre-industrial levels will mean "rapid, far-reaching and unprecedented changes in all aspects of society"."(3) The 3 year project by the IPCC is riddled with observations backed up strongly by data. The data is testable and peer reviewed. It does not matter your opinions on climate change the data is real and the implications are dire. Scientists on the panel say, with demonstrable evidence, that toying around a 1.5C rise starts to interfere with the livability and survivability of life on the planet. The previous thought was that if we could stay below 1.5C this century then the changes we would experience would be a little more manageable. New evidence though seems to shatter that hope. We seem to be barreling towards a 1.5C increase by the year 2030, a measly 12 years away!

What does that mean? Well once we start pushing past 1.5C the consequences start a chain reaction of devastation. At 2C the coral reefs will be completely dead. The oceans will be expected to rise by an additional 4 inches forcing mass migrations from coastal cities. If it continues its rise deserts will expand eradicating millions of acres of arable land that currently grow much of the world's food. Migration inland from the coastal cities will then turn into mass migrations north. The conditions to grow wheat and rice, two giant staples of the global diet, will likely not exist in anywhere near their current state. Ocean acidity would change quicker than sea life would be able to realistically adapt leading up to massive marine life extinctions. Society, nations, borders would not possibly be able to exist as we currently understand them.

In order to fully grasp what this report is telling us we must read the words from the report itself. I have been pouring over the actual report for a number of hours and a few things that stand out will be directly quoted here so as the reader has no excuse to mistake the words of distinguished and prominent scientists that compose the panel with the words of myself or a journalist. The report states directly:

"A1. Human activities are estimated to have caused approximately 1.0°C of global warming above pre-industrial levels, with a likely range of 0.8°C to 1.2°C. Global warming is likely to reach 1.5°C between 2030 and 2052 if it continues to increase at the current rate (high confidence).

B3. On land, impacts on biodiversity and ecosystems, including species loss and extinction, are projected to be lower at 1.5°C of global warming compared to 2°C. Limiting global warming to 1.5°C compared to 2°C is projected to lower the impacts on terrestrial, freshwater, and coastal ecosystems and to retain more of their services to humans (high confidence).

B5. Climate-related risks to health, livelihoods, food security, water supply, human security, and economic growth are projected to increase with global warming of 1.5°C and increase further with 2°C.

"With more than 6,000 scientific references cited and the dedicated contribution of thousands of expert and government reviewers worldwide, this important report testifies to the breadth and policy relevance of the IPCC," said Hoesung Lee, Chair of the IPCC.

Ninety-one authors and review editors from 40 countries prepared the IPCC report in response to an invitation from the United Nations Framework Convention on Climate Change (UNFCCC) when it adopted the Paris Agreement in 2015."(4)(5)

In an article by The New York Times titled, *Major Climate Report Describes a Strong Risk of Crisis as Early as 2040* the quote is as follows, "The report "is quite a shock, and quite concerning," said Bill Hare, an author of previous I.P.C.C. reports and a physicist with Climate Analytics, a nonprofit organization. "We were not aware of this just a few years ago."" (6)

If all of this has not been alarming enough this is where it starts to look even worse. In order to keep temperature rise below 1.5C a few key things must happen. Global CO_2 emissions must decrease by 45% from 2010 levels before we reach 2030. In order to do that industry will need to be all but eliminated. A 2017 *Carbon Majors* Report claims(with mountains of evidence to back it up) that 100 companies are responsible for over 70% of global emissions since 1988.(7) Multibillion dollar companies will sink the planet before they lose profits and the threat of them doing so is as real as it has ever been.

Also, renewable energy needs to supply 85% of all electricity by the year 2050. This is also a gigantic obstacle because of the immense wealth and power held by the fossil fuel industry. Profit over people is the staple of the capitalist society that reigns over the planet presently. Competition is heralded as a benefit in a capitalist system. When it comes to big oil fighting renewable energy, competition and a free market could be the final nail in the coffin of a planet whose economic system requires infinite growth within a system that has finite resources.

Coal use would need to reach nearly zero. Coal is not clean, coal is not sustainable. In order to have a chance at pushing back against 1.5C this is an absolute must in tandem with the previous two points. As if these three tasks where not monumental enough we would also need an additional seven million square kilometers of land to grow "energy crops" on. An energy crop would be categorized as a "plant grown at low cost and low maintenance harvest used to make biofuels or combusted for its energy content to generate electricity or heat." A land mass the size of Australia would be needed to grow these crops on while simultaneously we would need to see a global net emissions of zero by 2050.

If these things were implemented society as we know it would cease to exist. If nothing is done society as we know it would also cease to exist. There presently is no option C that is in the realm of possibility.

If that was not enough it gets worse. In August of 2018 the Trump administration released a 500-page report and buried deep inside that report was something that will knot the stomach if one can fully grasp what it means. Usually when issues of climate change are debated the side of the political spectrum that Trump resides on overwhelmingly either refuses to accept climate change is real, or to accept that it is being caused by man. The gut churning position the Trump administration took in its report is that "On its current course, the planet will warm a disastrous 7 degrees by the end of this century." (8) This is according to an article by the Chicago Tribune and Washington Post titled *Trump administration predicts a 7-degree global temperature rise. Its policies assume planet's fate is sealed.* One would hope the Trump administration finally chose to accept science and recognize and combat the clear and present danger to our planet. Quite the opposite actually, the report is used to justify the administration's policy of freezing a federal fuel efficiency standard. This decision would add more greenhouse gasses to the environment and the impact statement in the Trump administrations report says, "that policy would add just a very small drop to a very big, hot bucket." (8) The Trump administration is effectively saying the planet is on its way to catastrophe and it is unavoidable. The White House refused to offer comment.

This action makes the US government and the Trump administration a clear and immediate threat to human existence. The motivations for this decision piggy back on the reasons why it is unlikely we will be able to turn the situation around in time. Let's explore what I am saying clearly and completely.

The most likely course this slow-motion train wreck will manifest as follows. Anyone over the age of sixty can effectively distance themselves emotionally from this issue for the simple fact that they most likely will not be alive to see the worst of it. It is a basic human psychological tendency to be able to empathize with someone while simultaneously not actually taking action to help someone. I am not talking about your grandmother here. I am referring to the political and economic elite that have unarguable control over world economies and nations. The extremely wealthy are overwhelmingly male and they are overwhelmingly white. They are usually over half a century in age and they are living in luxury. To expect them to get behind dismantling the system that allows them to own five houses, three yachts, ten cars, and a private jet while 3.5 million children on this planet will starve to death at the same time this year is delusional thinking. Power never gives power away. Power only concedes some power when a threat of losing total power is imminent. So, when presented with the option of living out their remaining years in the classic bourgeois extravagant style or tearing down the system that serves them in favor of the greater good of humanity, you know which side of the proverbial fence they will come down on. Human behavior and history show us they will come down in favor of serving themselves every time. The political and economic sectors globally prop each other up and will continue to do so as we make the brisk march to the brink of existence.

The evidence is clear. Capitalism is going to kill this planet before we can grow out of the thinking that a global system predicated on production and destruction for profits sake is the one true destiny of human society. Karl Marx may end up being truly prophetic in saying that capitalism contains the seeds of its own destruction.

The IPCC climate report says that society needs to turn on a dime to avert this crisis. The reality that we exist in seems to point to us heading the opposite direction. We are having trouble and will continue to have trouble collectively grappling with the actual meaning of what we have now learned these many words below the opening sentence. The ever-increasing likelihood that we are careening toward, what could ultimately be, total human extinction is exacerbated by the fact that our desire to distract ourselves from the capitalistic imposed prison that surrounds us will only feed into our apathy. Apathy and inaction will not allow us to collectively divert this course and it is possible that my generation will be alive to die beside their children as the story of life on earth fades away into a footnote in the story of the universe.

Hold up, fuck that. We are not going out that easy. Here is what needs to happen. Religion needs to be thrown into the scrap heap of history for starters. In regard to reality we are not even sure what this all is. There are brilliant scientific minds who have presented evidence that we could be living in a mathematical simulation, while other great minds argue the opposite. Einstein's theories turn what we once knew as reality on its head. Theories exist dealing with multiverses and infinite reoccurring universes, these theories change and adapt with new understandings of physics and quantum mechanics. We have learned so much about the universe and what we truly are during our time on this rock and that's not even a fraction of a percent of what there still is to discover. So, I don't know much, but what I do know is that a bunch of prescientific ancient people didn't have the true nature of our existence figured out two thousand years ago. All of our advancements as a global society come from science not religion. Religion just ends up diddling your kids in the west and cutting off young girls' clitorises in the east. Religion claims to have the why and the who without being able to begin to fathom the how. It burned revolutionary minds at the stake in the dark ages and helped protect and smuggle many of those responsible for the holocaust out of Europe after World War II.

The main and most dangerous part about how religion manifests itself present day and especially within this specific topic is chilling. Religion may very well accept that the end of the world is coming, the problem is religion welcomes it. Islam, Christianity, Judaism they are all the chosen ones and they make up the majority of earths population. When God sorts the righteous out from the wicked it will be the believer's true paradise. The Christian rapture allows millions upon millions to indulge upon the idea of welcoming humanities destruction with the same geocentric era ideas that earths destruction means their salvation.

Religion must go but we must also stomach the idea that all the evidence points to the fact that industry must go as well. Transportation would need to undertake a major overhaul. The window that this has to happen in is extremely small and the margin for error is a lot smaller than we can really grasp. Coal and Capitalism need to share a rope and hang themselves if we are able to develop a sustainable sensible system in a timeframe that allows us to come out the other end of this lose lose situation.

Capitalism itself will eventually fall due to the inherent flaws that are not addressed. The amount of debt created with fiat currencies and bloated budgets becomes just one part of the inevitable downfall. The debt gets moved and transferred. We saw this on a "small scale" with the housing crisis of 2008 and we will see it on a larger scale some point in the future.

Eventually societal and planetary conditions with create a sense of unease and unrest and those that are owed money will be forced to call in their debts large and small in a short period. Since there is more global debt than actual currency by a large margin governments and corporations alike are going to be left holding the debt. Like one global game of hot debt potato. This is an essay not a textbook, so we will save the other reasons capitalism will implode on itself for another day.

The window is closing, the odds are widening, tank is empty. We got one shot to get it right, but human history has shown us we probably won't. This text is not a call to action it is a pre-death eulogy for a planet that was kind enough to allow us the conditions to thrive in only to have us turn around and plunge a spear right through its warm molten core. Humanity will be left standing like a deer in headlights hoping someone rises to lead them through this before they realize the solution is to come together and lead themselves. Are we really going to allow a bunch of greedy, power hungry, self-serving snakes do in this entire planet?

Since it is October 9th it is only fitting to end this essay with a quote by a man who was assassinated 51 years ago today by the CIA trained Bolivian military.

"Every day you have to fight so that love for humanity can be transformed into concrete deeds, into acts that set an example, that mobilize." -Ernesto "Che" Guevara, June 14, 1928 – October 9, 1967.

Figure it out, no one is going to save you. Work together and save yourselves. Or don't either way we are in for a wild ride. LoL.

Seriously though.

Why are more people not concerned about this?!

Oh well, my back hurts I'm about to go smoke a cig.

Ok im back.

J/k im out

Thanks for reading tho btw. Mad cool of u.

Don't @ me.

Citations

1. https://climate.nasa.gov/scientific-consensus/
2. Nuccitelli, Dana. Is the climate consensus 97%, 99.9%, or is plate tectonics a hoax? May 2017. https://www.theguardian.com/environment/climate-consensus-97-per-cent/2017/may/03/is-the-climate-consensus-97-999-or-is-plate-tectonics-a-hoax
3. McGrath, Matt. *Final call to save the world from 'climate catastrophe'* Oct 2018

https://www.bbc.com/news/science-environment-45775309?SThisFB

4. Global Warming of 1.5 °C *an IPCC special report on the impacts of global warming of 1.5 °C above pre-industrial levels and related global greenhouse gas emission pathways, in the context of strengthening the global response to the threat of climate change, sustainable development, and efforts to eradicate poverty* http://www.ipcc.ch/report/sr15/

5.

http://www.ipcc.ch/pdf/session48/pr_181008_P48_spm_en.pdf

6. Davenport, Coral. The New York Times. Oct 2018. *Major Climate Report Describes a Strong Risk of Crisis as Early as 2040*

https://www.nytimes.com/2018/10/07/climate/ipcc-climate-report-2040.html?login=email&auth=login-email

7. *The Carbon Majors Database: CDP Carbon Majors Report 2017* https://b8f65cb373b1b7b15feb-c70d8ead6ced550b4d987d7c03fcdd1d.ssl.cf3.rackcdn.com/cms/reports/documents/000/002/327/original/Carbon-Majors-Report-2017.pdf?1499691240

8. Eilperin, Juliet. Dennis, Brady. Mooney, Chris. *Trump administration predicts a 7-degree global temperature rise. Its policies assume planet's fate is sealed.* Sept 2018.
http://www.chicagotribune.com/news/nationworld/science/ct-trump-administration-global-warming-20180928-story,amp.html

www.ingramcontent.com/pod-product-compliance
Lightning Source LLC
Chambersburg PA
CBHW031524210526
45464CB00007B/3022